WITHDRAWN FROM
SAN BENITO COUNTY
FREE LIBRARY
NOT FOR RESALE

Animal Fathers

by RUSSELL FREEDMAN

drawings by JOSEPH CELLINI

Holiday House · New York

Text copyright © 1976 by Russell Freedman
Illustrations copyright © 1976 by Joseph Cellini
All rights reserved
Printed in the United States of America

Library of Congress Cataloging in Publication Data

Freedman, Russell.
   Animal fathers.

   SUMMARY: Discusses the child care behavior of fifteen animal fathers including the seahorse, Darwin's frog, emperor penguin, and gray wolf.

   1. Parental behavior in animals—Juvenile literature. [1. Parental behavior in animals. 2. Animals—Habits and behavior] I. Cellini, Joseph. II. Title.
QL762.F7   596′.05′6   75–35593
ISBN 0–8234–0273–8

Contents

COMMON AMERICAN SEAHORSE (*Hippocampus hudsonius*)	5
THREE-SPINED STICKLEBACK (*Gasterosteus aculeatus*)	8
SIAMESE FIGHTING FISH (*Betta splendens*)	10
SEA CATFISH (*Galeichthys felis*)	12

•

SMITH FROG (*Hyla faber*)	13
DARWIN'S FROG (*Rhinoderma darwinii*)	14
MIDWIFE TOAD (*Alytes obstetricians*)	16
TWO-TONED POISON-ARROW FROG (*Phyllobates bicolor*)	17

•

DOWNY WOODPECKER (*Dendrocopus pubescens*)	18
HOUSE WREN (*Troglodytes aedon*)	20
GREATER RHEA (*Rhea americana*)	22
EMPEROR PENGUIN (*Aptenodytes patagonica*)	24

•

GRAY WOLF (*Canis lupus*)	26
COMMON MARMOSET (*Callithrix jacchus*)	28
WHITE-HANDED GIBBON (*Hylobates lar*)	30

Most people think that baby animals are cared for mainly by their mothers. But this is not always the case. Many young animals cannot survive without the help and protection of their fathers.

Some animal fathers guard their babies and bring them food. Some build nests as nurseries for their young. Others carry their babies on their backs or in their bodies.

Often the father and mother work together as they watch over their young. Sometimes the father raises the family all by himself.

This book describes fifteen of the most interesting animal fathers.

Common American Seahorse

The seahorse is a curious fish. He has a head like a horse, a tail like a sea monster, and a pouch on his belly like a kangaroo. This pouch is where he raises his family.

When a mother seahorse is ready to lay her eggs, she swims close to the father. A small tube comes out of her body. She pushes this tube through a tiny opening in the father's pouch. Then she squirts a stream of eggs into the pouch. After that, she turns around and swims away.

The eggs split open inside the waterproof pouch. The baby seahorses are not yet fully developed, so they stay where they are. As they grow bigger, their father's pouch swells like a balloon. He carries his heavy load for six weeks before his babies are ready to be born.

When the time comes, the father coils his tail around a strand of seaweed. He bends sharply forward, then backward. He tightens the muscles in his pouch, trying to force the babies out. Sometimes he presses his full pouch against rocks and shells. At last he begins to give birth. Each time he tightens his muscles, a little seahorse is shot out into the water. As many as 150 babies float all around their father. When the last one is born, the father seahorse seems exhausted. His pouch has collapsed, and his children have scattered into the sea.

Three-Spined Stickleback

A stickleback is no bigger than your little finger. Often he spends a week or longer working on a nest for his young.

First he digs a pit at the bottom of a pond by scooping up sand with his mouth. Next he fills the pit with water weeds. He squirts the weeds with a sticky fluid from his body. Then he rubs against them, forming a neat pile. Finally he tunnels his way through the middle of the pile.

As the stickleback works, his belly turns bright red. This shows that he is ready to mate. When a female sees his red belly, she swims closer. He leads her to his nest and points to the tunnel with his nose. She squeezes into the tunnel. After she lays her eggs, the male chases her away. He stays behind to guard the eggs himself.

With his tail and fins, the father fans fresh water through the tunnel. He repairs parts of the nest that break off and float away. And he guards the nest from other fish. When the eggs hatch, the stickleback tears off the top of his nest. He leaves the bottom part as a cradle for the fry. If they roam too far, their father picks them up in his mouth and carries them home. They stay close to the nest for a few more days before swimming off on their own.

Siamese Fighting Fish

A Siamese fighting fish builds a floating nest of bubbles. He swims to the surface of his pond and sucks in some air. Then he spits out bubbles coated with saliva. Soon hundreds of tough, sticky bubbles are clustered together in the water.

When the male fighting fish sees a female, his whole body glows. Both of them swim in circles beneath the floating nest. As the mother sheds her eggs, they sink toward the bottom like tiny glass beads. The father races after them. He catches the eggs in his mouth and carries them back to his nest.

The mother swims away. The father fighting fish guards his nest. He attacks any fish that comes too close. If an egg falls out of the nest, he catches it with his mouth and puts it back. When the babies hatch, they hide among the bubbles for about a week. If they wander away, their father goes after them. He sucks the infants into his mouth and spits them back into their bed of bubbles.

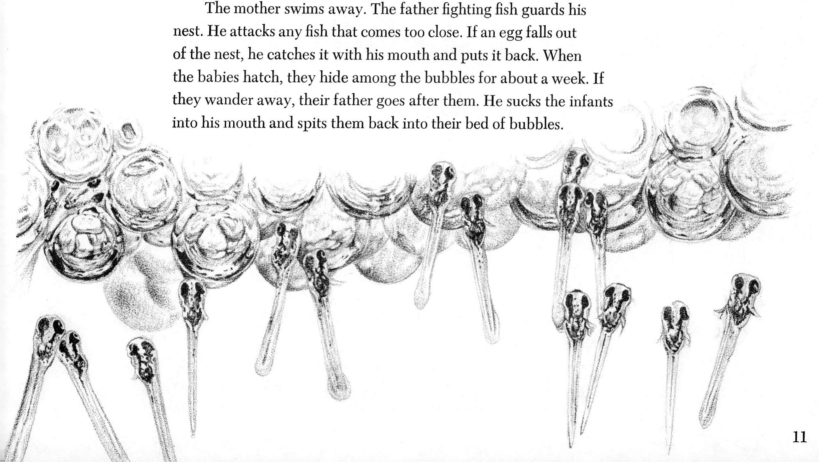

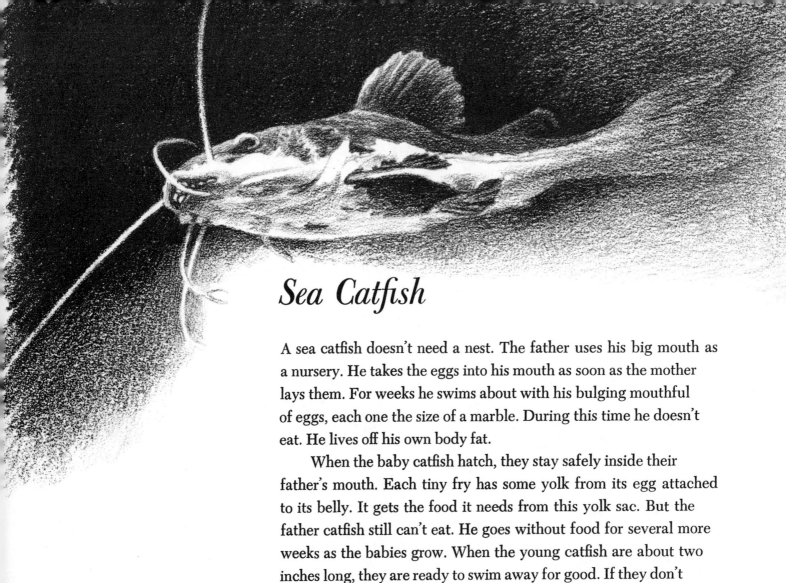

Sea Catfish

A sea catfish doesn't need a nest. The father uses his big mouth as a nursery. He takes the eggs into his mouth as soon as the mother lays them. For weeks he swims about with his bulging mouthful of eggs, each one the size of a marble. During this time he doesn't eat. He lives off his own body fat.

When the baby catfish hatch, they stay safely inside their father's mouth. Each tiny fry has some yolk from its egg attached to its belly. It gets the food it needs from this yolk sac. But the father catfish still can't eat. He goes without food for several more weeks as the babies grow. When the young catfish are about two inches long, they are ready to swim away for good. If they don't leave at the right time, their hungry father might swallow them.

Smith Frog

The South American smith frog is one of the few frogs that build nests. He climbs into the water at the edge of a pond and turns in a circle until he makes a hole in the mud. He builds a wall around the hole by pushing up mud with his snout. He pats the mud into place, making the wall smooth and strong. Inside the wall there is a sheltered pool for his eggs and young.

Now the smith frog calls loudly for a mate. A female comes along and lays her eggs in the water-filled nest. Both parents guard the eggs. When the tadpoles hatch, they grow up safely in their private swimming pool. The walls of the pool protect them from enemies in the rest of the pond. When they change from tadpoles into little frogs, they leave their nursery by climbing over the wall.

Darwin's Frog

Darwin's frog is no bigger than a cricket. And he is just as lively as he hops through forests in Chile and Argentina. He is the only frog anywhere that carries living tadpoles inside his body.

A female Darwin's frog lays from twenty to thirty eggs. Several males crowd around to guard the eggs. When the tadpoles are almost ready to hatch, they begin to squirm. Their jelly-covered eggs quiver and shake. Now the male frogs jump into action. Each male leaps forward and snaps up some eggs with his tongue. But he doesn't eat the eggs. Instead, he slides them through little slits in the floor of his mouth. The eggs drop into a long pouch inside his body, which stretches from his chin to his hips.

After the tadpoles hatch, they stay safely inside their father's elastic pouch. He goes about his usual business. His bulging pouch doesn't keep him from eating. Gradually the tadpoles lose their tails and develop into tiny frogs. Finally they leave their moist dark nursery. One by one, they jump out of the father frog's mouth.

Midwife Toad

Midwife toads live in Belgium and France. When the mother midwife lays her eggs, they are tied together like strings of beads. The father climbs onto the mother's back. He pushes his legs through the strings of eggs. He twists and squirms until the eggs are tangled around his thighs. Then the mother hops away, leaving the father in charge of the eggs.

During the day he hides in a burrow. At night he comes out to hunt for insects and worms. As he crawls through grass and weeds, the eggs on his legs are moistened by dew. This keeps them from drying out. Sometimes the father gives the eggs a good soaking in a puddle. In about a month the eggs are ready to hatch. The father goes to a pond and holds his legs underwater. The tadpoles break out of their eggs, wriggle their tails, and swim away.

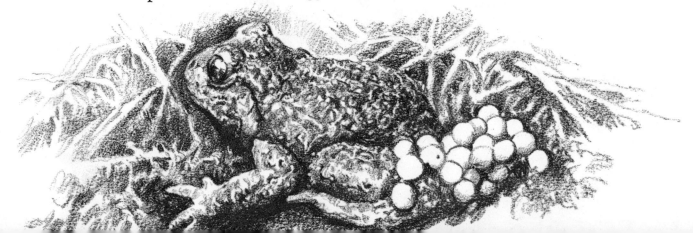

Two-Toned Poison-Arrow Frog

Poison-arrow frogs live in the jungles of South America. The mother frog lays her eggs on the ground. The father guards the eggs. When the tadpoles hatch, they wriggle onto their father's sticky back. He carries them piggyback through the jungle as they hang on with their tiny sucker-like mouths.

Every so often the father dunks his tadpoles in a pool to keep their skins moist. They cling tightly to his back, since they are not yet strong enough to swim. In a few weeks they are ready to fend for themselves. One day, as their father enters the water, the tadpoles drop off his back and go their own way.

Downy Woodpecker

A bird's nest may be built by the mother, the father, or both of them. Woodpecker parents work together. The father clings to the bark of a tree. He braces himself with his stiff tail and bangs away with his pointed beak. Then the mother takes over and drills for a while. They switch back and forth as they chop out their nesting hole.

Inside the nest, the mother lays her glossy white eggs on a bed of wood chips. Both woodpeckers take turns sitting on the eggs. And both of them care for the newly hatched chicks.

The mother stays home while the father hunts for insects and grubs. When he returns with a mouthful of food, he hammers loudly on the tree trunk. The mother peeks out to see who is knocking. They wave their heads in a greeting. Then the mother climbs out of the nest. The father squeezes inside to feed the chicks. Now he stays at home while the mother hunts for food.

In about a week, the chicks are big enough to stick their heads outside the nest at feeding time. In three weeks they are ready to fly.

House Wren

One nest isn't enough for a father house wren. When he flies north in the spring, he starts building several nests. First he searches for nesting holes. He pokes his head into an abandoned woodpeckers' nest, a hole in a wall, an old tin can, and an empty birdhouse. He picks out two or three holes and stuffs them with sticks and twigs.

Soon his bubbling song attracts a female. She inspects each of the male's sample nests. Finally she chooses one. She arranges the twigs to suit herself and lines the nest with feathers. Then she sits on her brown-spotted eggs while the father guards the nest.

The father and mother both work hard to feed their chicks. They fly back and forth all day long, bringing insects to their hungry brood. The chicks gulp down two hundred meals a day or more. In a month they are big enough to find their own food.

Now the father wren gets busy again. He cleans out the old nest and starts building new ones. He helps raise two or three broods in a single summer. Then he flies south for a long rest.

Greater Rhea

The greater rhea weighs eighty pounds and stands five feet tall. He can't fly, but he can run as fast as a horse.

These big gray birds live in Brazil and Argentina. The father rhea takes care of his young without help from anyone. He mates with several females. Then he tramples down the grass and scratches out a nesting hole. Each female leaves her eggs in his nest. Soon the nest is filled with twenty or thirty yellow eggs. They weigh two pounds apiece.

The father sits on the eggs for five weeks. He hisses loudly at any animal that comes too close, including the mother rheas. When the chicks hatch, they follow their father as he pecks at insects and leaves. He keeps a close watch on his big family. If he spots a fox or hawk, he warns the chicks with his booming voice. He will attack any animal that threatens the chicks. He may even attack a cowboy on horseback, or a small airplane landing nearby.

At night and when it rains, the father kneels on the ground. He gathers his chicks beneath his outspread wings. Some crawl up his back and nestle snugly in his feathers. Their father watches over them for three or four months until they are big enough to defend themselves.

Emperor Penguin

Emperor penguins live in frozen Antarctica, at the bottom of the world. They can't build nests because their home is covered by a vast sheet of ice.

A mother penguin lays one egg at the beginning of winter. Right away, the father rolls the egg on top of his feet. He covers the egg with a feathered flap of skin which hangs from his belly. For two months he stands shoulder to shoulder in a huge crowd of other father penguins. The temperature drops to forty degrees below zero. The coldest winds on earth howl around him at one hundred miles an hour. But his sagging belly keeps the egg warm.

When the egg hatches, the chick huddles between its father's feet. He bends down to feed the chick. It sticks its head into father's mouth and gurgles a milky liquid from his throat.

All of this time the father eats nothing at all. The mother penguins are miles away, swimming in the sea at the edge of the ice. They grow fat feeding on fish and squid. When the mothers return, the fathers are lean and hungry. Each father has lost about twenty-five pounds. At last the mothers take over. They slide in to feed the chicks with seafood stored in their bodies. Now the father penguins get their chance to head for the sea and start eating.

Gray Wolf

A father wolf doesn't get to see his puppies until they are about two weeks old. Newborn pups are blind and helpless. They stay with their mother inside her den, nursing and sleeping. Their father sleeps outside with the other wolves. He guards the entrance to the den. When he goes hunting, he brings food home to the mother.

The father enters the den for the first time when the pups are just beginning to see. He sniffs them and licks their fur and nips them gently. Then he rolls on his back and lets the puppies climb all over him.

Soon the pups are romping and playing outside. If a pup wanders away, the mother picks it up in her mouth and carries it back to the den. The father runs alongside, whining to the pup and licking its face. When the pups begin to gnaw on bones and gulp down meat, their father brings food home to them. Every wolf in the pack helps care for the puppies. All the wolves guard the youngsters, feed them and play with them.

In about two months the pups are big enough to live in the open. From then on they travel with the pack.

Common Marmoset

Marmosets are tiny South American monkeys about the size of squirrels. The rain forests echo with their shrill cries as they scurry through trees and crouch on branches.

    Baby marmosets depend on their fathers. A mother marmoset usually has twins or triplets. As the babies are born, the father takes them and cleans their fur. From then on he carries the helpless infants through the forest. They cling tightly to their father's back as he leaps from branch to branch. He returns the babies to their mother only at feeding time. She holds them as they nurse. Then she hands them back to the father.

When the babies are ready for solid food, their father helps feed them. He picks fruit and catches insects. He chews this food himself before giving it to the young ones.

In a few weeks the youngsters are big enough to ride on their father's shoulders. Soon they make their first timid attempts to climb trees. If they stumble or fall, their father rushes over to rescue them. By the time the young marmosets are four months old, they can race through the treetops by themselves. But they still depend on their parents for protection. At night they sleep with their father and mother high above the forest floor.

White-Handed Gibbon

A father gibbon moves swiftly through the Thailand forest. He swings from branch to branch by his long arms. At times he leaps twenty or thirty feet from tree to tree. His family leaps and swings after him. A small baby clings to the mother's hip. An older sister and brother hoot and shout as they follow their parents.

All day long the father leads his family along treetop trails. He keeps them together with his loud calls. When he finds a tree filled with sweet ripe plums, he shouts to the others. As they feast on the purple fruit, they can hear the calls of neighboring gibbons. These small apes always live in family groups. Each family stays in its own part of the forest.

During the hottest part of the day, the gibbons stop to rest. The father and mother sit close together on a shady branch. The baby nurses in its mother's arms. The sister and brother chase each other in a wild game of tag. Suddenly the sister falls. She grabs a branch and cries for help. The father races to her side. He wraps his arms around the youngster and hugs her tightly.

As it grows dark, the father heads for a favorite sleeping tree. He picks out a sturdy branch, hidden among the leaves. Then he curls up beside the mother and baby. The sister and brother huddle together on a branch below. Their father and mother will look after them until they are six or seven years old.